The nature of science

Understanding what science is all about

Written by Dorothy Warren
RSC School Teacher Fellow 1999–2000

ROYAL SOCIETY OF CHEMISTRY

The nature of science

Written by Dorothy Warren

Edited by Colin Osborne and Maria Pack

Designed by Imogen Bertin

Published and distributed by Royal Society of Chemistry

Printed by Royal Society of Chemistry

For further information on other educational activities undertaken by the Royal Society of Chemistry write to:

Education Department
Royal Society of Chemistry
Burlington house
Piccadilly
London W1J 0BA

Information on other Royal Society of Chemistry activities can be found on its websites:
http://www.rsc.org
http://www.chemsoc.org
http://www.chemsoc.org/LearnNet contains resources for teachers and students from around the world.

ISBN 0–85404–376–4

British Library Cataloguing in Publication Data.

A catalogue for this book is available from the British Library.

RS•C

Foreword

Many students in school have a view of science that is far removed from the real world practice of science and they too often consider science to be a defined body of knowledge and scientists to know all the answers. This book tries to give students an awareness of the processes of science and of the nature of science as a changing body of knowledge with uncertainties and much remaining to be discovered. It is hoped that, by using it, teachers can communicate to their students some of the wonder and excitement of science and encourage the development of future generations of scientists.

Professor Steven Ley CChem FRSC FRS
President, The Royal Society of Chemistry

RS•C

Acknowledgements

The production of this book was only made possible because of the advice and assistance of a large number of people. To the following, and everyone who has been involved with this project, including the members of the science staff and students in trial schools, both the author and the Royal Society of Chemistry express their gratitude.

General

Colin Osbome, Education Manager, Schools & Colleges, Royal Society of Chemistry
Maria Pack, Assistant Education Manager, Schools & Colleges, Royal Society of Chemistry
Members of the Royal Society of Chemistry Committee for Schools and Colleges.
Members of University of York Science Education Group.
Jill Bancroft, Special educational needs project officer, CIEC
Donald Stewart, Dundee College, Dundee
Richard Warren, Mathematics Department, Ampleforth Colledge, York
Bob Campbell, Department of Educational Studies, University of York
Professor David Waddington, Department of Chemistry, University of York.
Elizabeth Pritchard, Laboratory of the Government Chemist
Peter Dawson, Science adviser, York and the York Schools Secondary science group.

Pictures

National Maritime Museum

Schools

David Billett, Ampleforth College, York.
Peter Bird, Alderwasley Hall School, Derbyshire.
Sandra Buchanan, Tobermory High School, Isle of Mull.
Howard Campion, Fulford School, York.
Carole Lowrie, Hummersknott School, Darlington
Susan Vaughan, All Saints School, York.
Gavin Cowley, Oaklands School, York

The Royal Society of Chemistry would like to extend its gratitude to the Department of Educational Studies at the University of York for providing office and laboratory accommodation for this Fellowship and the Head Teacher and Governors of Fulford Comprehensive School, York for seconding Dorothy Warren to the Society's Education Department.

RS•C

RS•C

Contents

RS•C

How to use this resource

At the start of the 21st century secondary education yet again underwent changes. These included the introduction of new curricula at all levels in England, Wales and Scotland and the Northern Ireland National Curriculum undergoing review. With more emphasis on cross curricula topics such as health, safety and risk, citizenship, education for sustainable development, key skills, literacy, numeracy and ICT, chemistry teachers must not only become more flexible and adaptable in their teaching approaches, but keep up to date with current scientific thinking. The major change to the science 11–16 curricula of England and Wales was the introduction of 'ideas and evidence in science', as part of Scientific Enquiry. This is similar to the 'developing informed attitudes' in the Scottish 5–14 Environmental studies, and is summarised in Figure 1.

In this series of resources, I have attempted to address the above challenges facing teachers, by providing:

- A wide range of teaching and learning activities, linking many of the cross-curricular themes to chemistry. Using a range of learning styles is an important teaching strategy because it ensures that no students are disadvantaged by always using approaches that do not suit them.

- Up-to-date background information for teachers on subjects such as global warming and Green Chemistry. In the world of climate change, air pollution and sustainable development resource material soon becomes dated as new data and scientific ideas emerge. To overcome this problem, the resources have been linked to relevant websites, making them only a click away from obtaining, for example, the latest UK ozone data or design of fuel cell.

- Resources to enable ideas and evidence in science to be taught within normal chemistry or science lessons. There is a need to combine experimental work with alternative strategies, if some of the concerns shown in Figure 1, such as social or political factors, are to be taught. This can be done for example, by looking at the way in which scientists past and present have carried out their work and how external factors such a political climate, war and public opinion, have impinged on it.

- Activities that will enhance student's investigative skills.

These activities are intended to make students think about how they carry out investigations and to encourage them to realise that science is not a black and white subject. The true nature of science is very creative, full of uncertainties and data interpretation can and does lead to controversy and sometimes public outcry. Some of the experiments and activities will be very familiar, but the context in which they are embedded provide opportunities for meeting other requirements of the curriculum. Other activities are original and will have to be tried out and carefully thought through before being used in the classroom. Student activities have been trialled in a wide range of schools and where appropriate, subsequently modified in response to the feedback received.

Dorothy Warren

RS•C

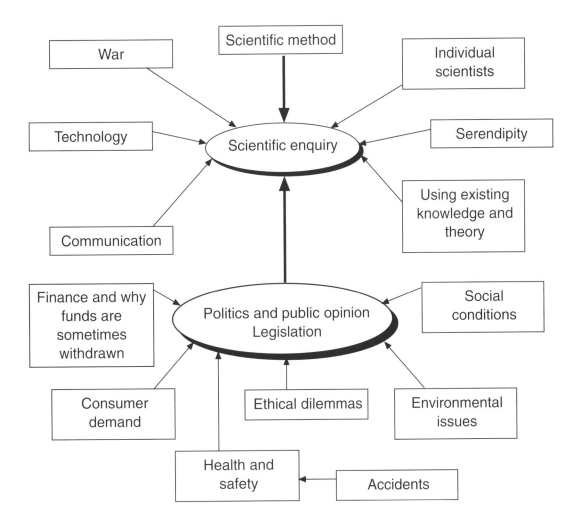

Figure 1 The factors influencing the nature of scientific ideas– scientific enquiry and the advancement of science

Maximising the potential use of this resource

It is hoped that this resource will be widely used in schools throughout the United Kingdom. However, as every teacher knows, difficulties can be experienced when using published material. No single worksheet can cater for the needs of every student in every class, let alone every student in every school. Therefore many teachers like to produce their own worksheets, tailored to meet the needs of their own students. It was not very surprising when feedback from trial schools requested differentiated worksheets to allow access to students of different abilities. In an attempt to address these issues and concerns, this publication allows the worksheet text and some diagrams to be modified. All the student worksheets can be downloaded in Word format, from the Internet via the LearnNet website, **http://www.chemsoc.org/networks/learnnet/ideas-evidence.htm** .This means that the teacher can take the basic concepts of the activity, and then adapt the worksheet to meet the needs of their own students. Towards the end of the teachers' notes for most activities there are some suggestions as to how the resource can be adapted to meet the needs of students of different abilities. There are also some examples of differentiated worksheets included in the resource.

RS•C

It is not envisaged that teachers will use every activity from each piece of work with an individual class, but rather pick and choose what is appropriate. For example some activities use high level concepts and are designed to stretch the most able student and should not be used with less able students *eg* Interpreting the weighing experiment methods 1 and 2.

Activities that involve researching for secondary information on the Internet contain hyperlinks to appropriate websites. To minimise the mechanical typing of the URLs and possible subsequent errors, the students can be given the worksheet in electronic form and asked to type in their answers. The websites are then only a click away.

Appropriate secondary information has been included in the teachers' notes for use in class when the Internet or ICT room is unavailable.

Unfortunately, from time to time website addresses do change. At the time of publication all the addresses were correct and the date that the site was last accessed is given in brackets. To minimise the frustration experienced when this happens, it is advisable to check the links before the lesson. If you find that a site has moved, please email both **LearnNet@rsc.org** and **education@rsc.org** giving full details so that the link can be updated on the worksheets on the web in the future.

Strategies for differentiated teaching

All students require differentiated teaching and it is not just an issue for those students with special educational needs. The following definition by Lewis[1] has been found to be quite useful.

'Differentiation is the process of adjusting teaching to meet the needs of individual students.'

Differentiation is a complex issue and is very hard to get right. It can be involved in every stage of the lesson *ie* during planning (differentiation by task), at the end of the activity (differentiation by outcome) and ongoing during the activity. Often teachers modify the activity during the lesson in response to feedback from the class. Differentiation does not only rely on appropriate curriculum material but is also concerned with maximizing learning. Student involvement and motivation effect the learning experience and should be considered and taken into account. It is therefore not surprising that differentiation is one of the areas of classroom teaching where teachers often feel under-confident. Most strategies for differentiated lessons are just applying good teaching practice eg varying the pace of the lesson, providing suitable resources and varying the amount and nature of teacher intervention and time.[2] Rather than just providing several examples of differentiated activities from the same worksheet, a list of strategies for differentiated teaching is presented, with some examples of how they can be used in the classroom. The examples can be found at the appropriate places in the text.

1. Using a range of teaching styles

A class is made up of different personalities, who probably have preferred learning styles. Using a range of teaching approaches makes it more likely that all students will be able to respond to the science that is being taught. The following examples have been included and can be found at the appropriate place in the resource.

Example Scurvy – the mystery disease
Approach 1 – A paper exercise analysing the James Lind experiment
Approach 2 – James Lind role-play

RS•C

2. Varying the method of presentation or recording
Giving the students some choice about how they do their work. There are many opportunities given throughout the resource.

3. Taking the pupil's ideas into account
Provide opportunities for students to contribute their own ideas to the lesson. For example when setting up an investigation allow different students the freedom to chose which variables they are going to investigate. The use of concept cartoons provides an ideal opportunity for students to discuss different scientific concepts. **Flickering candles** and **Brewing up** both set the scene for going on to investigate combustion. Essentially the students will all require the same equipment, but they may choose to investigate slightly different questions.

4. Preparing suitable questions in advance
Class discussions are important in motivation, exploring ideas, assessment etc. Having a list of questions of different levels prepared in advance can help to push the class.

5. Adjusting the level of scientific skills required
Example – Using symbol equations or word equations

6. Adjusting the level of linguistic skills required
Teachers may like to check the readability of their materials and of the texts they use. Guidance on this and on the readability of a range of current texts may be found at **http://www.timetabler.com/contents.html** (accessed June 2001).

References

1. A. Lewis, *British Journal of Special Education*, 1992, **19**, 24–7.

2. S. Naylor, B. Keogh, *School Science Review*, 1995, **77(279)**, 106–110.

How scientists communicate their ideas

Effective communication is crucial to the advancement of science and technology. All around the globe there are groups of research scientists and engineers, in universities and in industry, working on similar scientific and technological projects. Communication between these groups not only gives the scientist new ideas for further investigations, but helps in the evaluation of data. Results from different groups will either help to confirm or reject a set of experimental data. Communication is vital when a company wants to sell a new product. Depending on the product the buyer will want to understand how it works and how to maintain it. Several of the employees will have to learn how to use the product, and respond quickly to changing technology and circumstances. Therefore the manufacturers must be able to communicate the science to prospective buyers.

Scientists communicate in a number of ways including:

■ Publication in research journals

■ Presenting papers at scientific conferences

■ Poster presentations at conferences

■ Book reviews by other scientists

■ Publication on the Internet

■ Sales brochures

RS•C

- Advertising flyers
- Television documentaries

Publication in research journals

The article is written. The article must have an abstract, which is a short summary.

It is submitted to a journal.

The article is refereed by other scientists, working in a similar area. This is to check that the work is correct and original.

The article may be returned to the author to make changes.

The article is accepted and published by the journal.

The article is published.

Presenting papers at scientific conferences

Conference organisers invite scientists to speak on specific topics and projects.

An abstract is submitted to and accepted by the conference organisers.

The conference programme is organised and the speakers notified.

The scientist gives their talks, usually aided by slides, which contain the main points.

There is usually time for questions after the talk.

The written paper is given to the conference organisers.

All the papers are published in the conference proceedings. This is usually a book.

Poster presentations at conferences

An abstract is submitted to and accepted by the conference organisers.

The conference programme is organised and the poster people notified.

During the poster session the authors stand by the posters, ready to answer any questions as the delegates read the posters.

Written papers may then be published in the conference proceedings.

Book reviews

Other scientists in the same field often review new books. The reviews are then published in scientific magazines and journals. The review offers a critical summary of the book. The idea of the review is to give possible readers an idea of the contents and whether it is suitable for the intended purpose.

Publishing on the Internet

This is the easiest way to publish. Anyone can create their own web page and publish their own work. In this case the work is not refereed or checked by other people.

However, a lot of the information published on the Internet is linked to reputable organisations. In this case the articles will have been checked before they are published. Much of the information published on the Internet is targeted at the general public, and therefore the scientific ideas are presented in a comprehensible way. There are often chat pages so people can communicate their views and ask questions or request further information. The power of the Internet is that there is the opportunity to get immediate feedback to a comment or question.

Sales brochures

The information must be presented in an attractive and concise manner. After all you are trying to sell something. There should be a balance between technical information and operating instructions!

Advertising flyers

This must be written with the target audience in mind.

The information must be concise as there is limited space. The format must be attractive and should include pictures as well as writing. The flyer should also be quite cheap to produce.

Teaching students to communicate ideas in science

Students can be taught effective communication skills:

■ By encouraging communication between students and a range of audiences in classrooms

■ By encouraging them to investigate like 'real scientists' by reporting their findings for checking and testing by others, and participating in two-way communication. (Communicating between groups, classes, partner schools, schools abroad perhaps via the Internet.)

■ By setting investigations in a social context which offers the opportunity to communicate the project outside of the classroom. These work best when there is local interest.

When presenting investigative work to an audience, the student should consider the following:

■ Who will be in the audience?

■ What information does the audience need to know eg method, results and recommendations?

■ How to present the information in an interesting and professional way eg should graphs be hand drawn or done on the computer?

■ That the information offered convinces the audience that their investigation was valid and reliable.

■ Poster presentations or display boards should be concise, since the space is limited.

■ When speaking to audiences remain calm, speak clearly and slowly and try to be enthusiastic. Make sure that information on slides and OHTs can be read from the back of the room.

When writing a report of the findings of a scientific investigation for others to check and test, the emphasis should be on clarity. Another person is going to carry out the same investigation. The only information available is what is written in the report.

The report could be written under the following headings:

■ Introduction

■ Scientific knowledge

■ Planning

■ Table of results

■ Graphs

RS•C

■ Conclusions

■ Evaluation

■ Recommendations.

Further background information

R. Feasy, J. Siraj-Blatchford, *Key Skills: Communication in Science*, Durham: The University of Durham / Tyneside TEC Limited, 1998.

Curriculum coverage

Curriculum links to activities in this resource are detailed at
http://www.chemsoc.org/networks/learnnet/nature.htm

Curriculum links to activities in other resources in this series are detailed at

http://www.chemsoc.org/networks/learnnet/ideas-evidence.htm

Health and safety

All the activities in this book can be carried out safely in schools. The hazards have been identified and any risks from them reduced to insignificant levels by the adoption of suitable control measures. However, we also think it is worth explaining the strategies we have adopted to reduce the risks in this way.

Regulations made under the Health and Safety at Work etc Act 1974 require a risk assessment to be carried out before hazardous chemicals are used or made, or a hazardous procedure is carried out. Risk assessment is your employers responsibility. The task of assessing risk in particular situations may well be delegated by the employer to the head of science/chemistry, who will be expected to operate within the employer's guidelines. Following guidance from the Health and Safety Executive most education employers have adopted various nationally available texts as the basis for their model risk assessments. These commonly include the following:

Safeguards in the School Laboratory, 11th edition, ASE, 2001

Topics in Safety, 3rd Edition, ASE, 2001

Hazcards, CLEAPSS, 1998 (or 1995)

Laboratory Handbook, CLEAPSS, 1997

Safety in Science Education, DfEE, HMSO, 1996

Hazardous Chemicals – a manual for science education, SSERC, 1997 (paper)

Hazardous Chemicals – an interactive manual for science education, SSERC, 1998 (CD-ROM)

If your employer has adopted more than one of these publications, you should follow the guidance given there, subject only to a need to check and consider whether minor modification is needed to deal with the special situation in your class/school. We believe that all the activities in this book are compatible with the model risk assessments listed above. However, teacher must still verify that what is proposed does conform with any code of practice produced by their employer. You also need to consider your local circumstances. Is your fume cupboard reliable? Are your students reliable?

Risk assessment involves answering two questions:

RS•C

■ How likely is it that something will go wrong?

■ How serious would it be if it did go wrong?

How likely it is that something will go wrong depends on who is doing it and what sort of training and experience they have had. In most of the publications listed above there are suggestions as to whether an activity should be a teacher demonstration only, or could be done by students of various ages. Your employer will probably expect you to follow this guidance.

Teachers tend to think of eye protection as the main control measure to prevent injury. In fact, personal protective equipment, such as goggles or safety spectacles, is meant to protect from the unexpected. If you expect a problem, more stringent controls are needed. A range of control measures may be adopted, the following being the most common. Use:

■ a less hazardous (substitute) chemical;

■ as small a quantity as possible;

■ as low a concentration as possible;

■ a fume cupboard; and

■ safety screens (more than one is usually needed, to protect both teacher and students).

The importance of lower concentrations is not always appreciated, but the following table, showing the hazard classification of a range of common solutions, should make the point.

Ammonia (aqueous)	irritant if ≥ 3 M	corrosive if ≥ 6 M
Sodium hydroxide	irritant if ≥ 0.05 M	corrosive if ≥ 0.5 M
Ethanoic (acetic) acid	irritant if ≥ 1.5 M	corrosive if ≥ 4 M

Throughout this resource, we make frequent reference to the need to wear eye protection. Undoubtedly, chemical splash goggles, to the European Standard EN 166 3 give the best protection but students are often reluctant to wear goggles. Safety spectacles give less protection, but may be adequate if nothing which is classed as corrosive or toxic is in use. Reference to the above table will show, therefore, that if sodium hydroxide is in use, it should be more dilute than 0.5M (M=mol dm^{-3}).

CLEAPSS Student Safety Sheets

In several of the student activities CLEAPSS student safety sheets are referred to and recommended for use in the activities. In other activities extracts from the CLEAPSS sheets have been reproduced with kind permission of Dr Peter Borrows, Director of the CLEAPSS School Science Service at Brunel University.

■ Teachers should note the following points about the CLEAPSS student safety sheets:

■ Extracts from more detailed student safety sheets have been reproduced.

■ Only a few examples from a much longer series of sheets have been reproduced.

■ The full series is only available to member or associate members of the CLEAPSS School Science Service.

RS•C

- At the time of writing, every LEA in England, Wales and Northern Ireland (except Middlesbrough) is a member, hence all their schools are members, as are the vast majority of independent schools, incorporated colleges and teacher training establishments and overseas establishments.

- Members should already have copies of the sheets in their schools.

- Members who cannot find their sheets and non-members interested in joining should contact the CLEAPSS School Science Service at Brunel University, Uxbridge, UB8 3PH; tel. 01895 251496; fax. 01895 814372; email science@cleapss.org.uk or visit the website **http://www.cleapss.org.uk** (accessed June 2001).

- In Scotland all education authorities, many independent schools, colleges and universities are members of the Scottish Schools Equipment Resource Centre (SSERC). Contact SSERC at St Mary's Building, 23 Holyrood Road, Edinburgh, EH8 8AE; tel. 0131 558 8180, fax 0131 558 8191, email sts@sserc.org.uk or visit the website **http://www.sserc.org.uk** (accessed June 2001).

RS•C

RS•C

Introduction

There are many different approaches to science. Included in this book are a range of activities designed to look at different aspects of the nature of science and to teach investigative skills. The scientific method has considerably developed over the last 500 years and **Scurvy – the mystery disease** looks at how the scientific method changed and developed during the 400 year quest to find a cure for the disease. The worksheet overleaf is just one approach that some scientists may use today when carrying out an investigation. It is important for students to understand that a theory or model often changes over time as new evidence is collected and that it is extremely important to check the accuracy and precision of the measurements.

A scientific theory or model is a simple or complex explanation put forward by scientists to explain various phenomena. A model is used to help scientists visualise things that they cannot actually see. The theory or model is usually based on previous scientific knowledge and experiences, as well as careful observation and measurements. The theory can then be used to explain further phenomena and to make predictions of future behaviour. Therefore scientific theories and models are powerful tools. However, they are only valid as long as they can be used to explain all the available data, *ie* from both observations and measurements. Scientists will often test out theories by carefully designing and carrying out experiments. If new data appears that does not fit the theory, then the theory may have to be modified and updated. It will then have to be tested out again. We could say that scientific theories and models are 'living' because they change and evolve. (Of course in practice, life is not usually that simple and other factors interfere.)

Students often find it difficult to distinguish between observation and inference, *ie* putting your own interpretation on an observation. Students often believe that scientific knowledge is provable in an absolute sense, and do not consider science as a creative subject relying to a certain extent on human imagination. It is also common for students to believe that laws are theories that have been proven, and once they are proved they will not change. It is also common for people to believe that there is only one route to the solving the answer. The **Black box activities** should help to dispel some of these myths. They are called Black box activities because the students are 'working in the dark' and have to base their inferences on observation.

Accuracy and precision plays an important role in scientific investigations from choosing the correct piece of measuring equipment, to knowing when to discard a data point as anomalous and how many significant figures or decimal places to quote in the final answer. Three student activities have been included to address some of the issues of accuracy and precision.

Experience shows that it is often difficult to get students to put forward their own scientific ideas when introducing an investigation. Concept cartoons are designed to overcome this problem by presenting the students with a range of possible answers to a scientific question. The concept cartoons can be used to promote discussion. Good discussions have many uses, for example they can help students to form their own scientific ideas, which they can use to plan their own investigations. **Brewing up** and **Flickering candles** are two examples of concept cartoons on the theme of combustion.

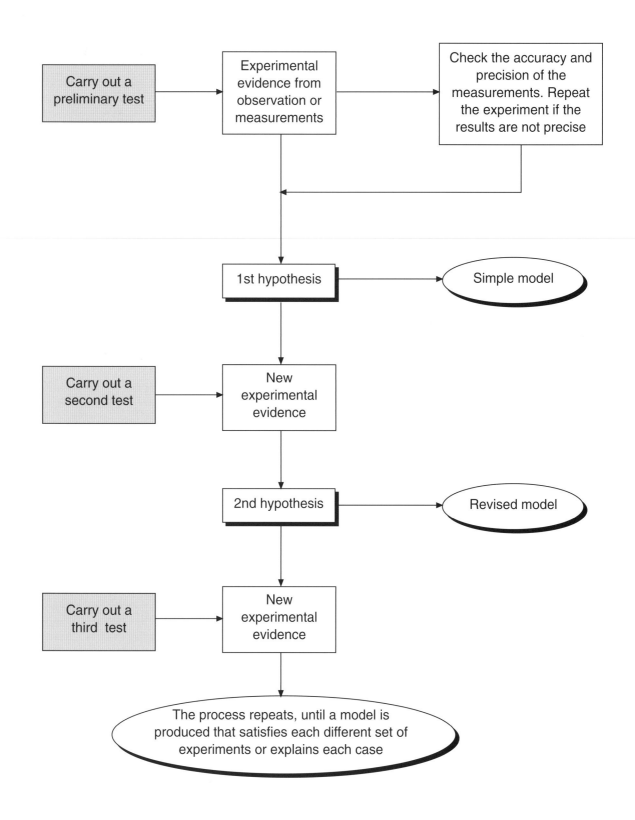

Making a model or theory

RS•C

Black box activity 1: Tricky tracks

Teachers' notes

Objectives

■ To help the students distinguish between observation and inference.

■ To introduce the concept that all ideas are valid unless there is further evidence to suggest otherwise.

Teaching topics

This activity is appropriate for 11–13 years olds and fits into a scheme of work anywhere where observations are made. It can also be successfully used with lower ability 14–15 year olds. A possible follow up activity would be to carry out a series of class 'test tube' experiments, which required careful observation, followed by interpretation.

Background information

This activity assumes the following definitions:

■ An observation is what is actually seen.

■ An inference is interpreting what you see.

Teaching tips

This activity could be carried out as a whole class exercise, with individual students recording their own answers or in groups of 3–4 students.

The lesson could be introduced as follows:

Scientists are always putting forward ideas or theories to try and explain the things they see happening in the world.

Scientists often test their ideas and theories by carrying out some experiments. Sometimes new ideas come along which do not fit the old ideas. So more experiments have to be carried out to see if the new idea is correct.

Resources

■ OHP – if the activity is carried out as a whole class exercise. Photocopiable worksheets **Tricky tracks 1–3** will need to be transferred onto OHT slides.

■ Class sets of student worksheets if the activity is to be carried out in groups
  – Tricky tracks 1
  – Tricky tracks 2
  – Tricky tracks 3
  – Questions

Timing

60–70 minutes. Approximately half the time should be used to carry out the worksheet tasks and the rest of the time discussing the answers.

RS•C

Possible lesson plan

1. Give out the student worksheet **Tricky tracks 1** and the **Making a model or theory** worksheet. The class starts the worksheets either individually or in groups and collects **Tricky tracks 2** and **3** as required.

   Or if using an OHP, put up **Tricky tracks 1** and give the class 10 minutes to do question 1, before putting up **Tricky tracks 2** and **Tricky tracks 3**.

2. Feedback to the whole class some answers to question 1, by asking different students to read out their accounts. Try and get as many different explanations as you can. It is important to accept all explanations equally.

3. Go through the rest of the questions, pointing out the difference between observation and inference.

4. Finally, link this exercise to scientists, saying that scientists often make similar inferences as they try to interpret their observations. There may be several equally valid theories until new evidence comes along to change it.

Next lesson
Carry out a class practical which involves recording observations and deductions in a table. For example, the reaction of metals and acids, or calcium and water, or iron and sulfur.

Adapting resources

The student worksheet can be easily adapted to meet the needs of less able students by omitting some of the more demanding questions, such as 3 and 7, and renumbering the other questions.

Some students may find question 3 confusing and may become frustrated by the exercise (see answers), but it should present no problems to the more able student.

Answers

1. All answers to question 1 should be accepted.

Possible answers to question 1 could include:

a) Two animals or birds were out hunting in the forest. On meeting each other at the clearing, they had a fight.

b) Two animals or birds had a fight in the forest and then went off home to different places.

c) The tracks lead down to the place where birds meet before going for a swim. During the day many different birds will come and go, that is why you can see so many tracks. The big birds and little birds live in different places and that is why the big tracks go off in a different direction to the small tracks. Most of the birds arrive and leave by air, so you cannot see their tracks on the ground. Usually the small birds and the large birds go to the ponds at different times because the little birds try to avoid the big birds, which often chase them.

d) One day a large male dinosaur was out looking for food. As he was searching a small area he noticed a beautiful female dinosaur coming towards him. It was love at first sight. Being a little shy, he hid behind a rock and when she got closer he jumped out and introduced himself. At first she was afraid, she thought that he was going to attack her and she tried to escape. But when he spoke and told her not to be afraid, they sat down on the rock and had a good chat.

2. In **Tricky tracks 2** we see two sets of marks. The marks on the left side of the page are bigger than on the right side of the page. Each big mark has three points coming out of a small black blob. The marks form a diagonal line. Each mark is pointing in the same direction.

The marks on the right hand side of the page have three points coming out of an elongated blob. There are two rows of marks going along the side of the page.

There are 14 big marks and 34 small marks.

Only accept answers that describe what is actually seen on the page.

Do not accept any answers that try to interpret what the marks are.

A typical student answer might be:
'I can see two sets of bird tracks on the page. One bird is bigger than the other bird.'

Challenge the student: How do you know they are bird tracks? Could the tracks belong to anything else?

Explanation: You should now explain that an observation is what you see on the page, not what you think you see on the page.

The typical student answer given above, had been interpreted by using some other knowledge.

The student reasoning could be as follows:

'From previous knowledge, I know that these marks resemble birds tracks. I know that big birds have bigger feet than small birds, therefore I have used bird tracks in my answer.'

Remind the class of the question, 'What do you observe in **Tricky tracks 2**?'

3. This is really a trick question. We do not know that the tracks have been made by animal and we do not know that they are going in the same direction. It should be pointed out that this question has been built on two assumptions. Often in science, we have to be careful about the questions we ask.

4 & 5 These should be treated in the same manner as question 2. You may wish to give the class time to change the answer to their questions, in the light of the answer to question 2.

6. Based on the current evidence, we can never know what has really happened. We can only imagine what has happened. Therefore, at each stage all the theories put forward in question 1 are equally valid.

7. To find out more about the situation we could look try and identify the tracks using a key. Once the tracks were identified, more could be found out about the behaviour of the animals, which may offer support to some theories and not to others.

Tricky tracks 1

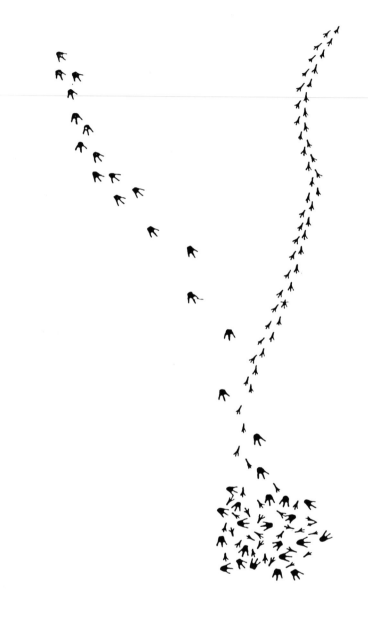

Tricky tracks 2

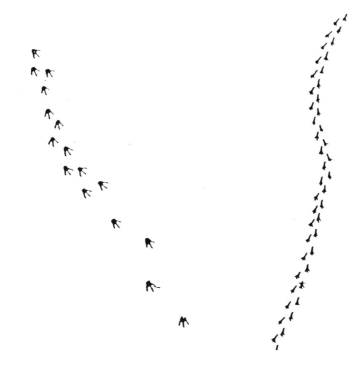

Tricky tracks 3

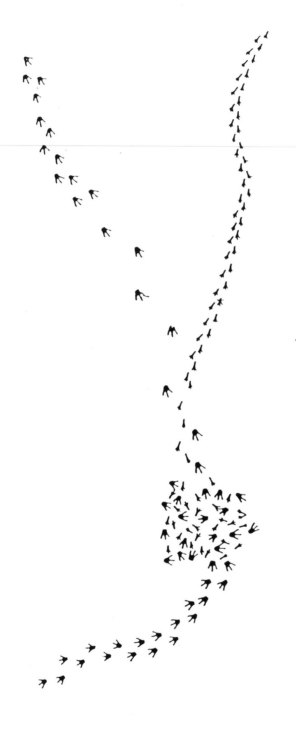

Questions

1. Carefully study **Tricky tracks 1** and write a short account of what you think has happened.

Now collect Tricky tracks 2 from your teacher

2. What do you observe in **Tricky tracks 2**?

3. Why are the two animals heading towards the same point?

Now collect Tricky tracks 3 from your teacher

4. What do you observe in **Tricky tracks 3**?

5. What do you observe in **Tricky tracks 1**?

6. Based on what you have found out so far, do you think we can ever know what has really happened?

7. What could you do to find out more about the situation?

RS•C

Black box activity 2: The cube activity

Teachers' notes

Objectives
■ To look for patterns and use them to work out the missing information.

Outline
The search for patterns based on data is a large part of the scientific method. The patterns can then be used to predict further data points and then are often applied to other systems. It should be emphasised that scientific knowledge is based partly on observation and experiment, and partly on human creativity in interpretation.

Teaching topics

This activity is appropriate for 11–12 year olds and can be easily fitted into a scheme of work. The exercise could be used to lead into some 'real data' interpretation exercises either using data gathered from their own experiments or data from secondary sources.

Teaching tips

This activity could be carried out in groups of 2–4.

Resources (for each group)

■ Copies of the templates for the three cubes or cubes already made up

■ Scissors

■ Sellotape

■ Card or paper (for making more cubes)

■ Student worksheets
– The cube activity

Timing

One lesson. It can be easily lengthened or shortened by varying the different number of cubes given to the class.

Possible lesson plan

1. Hand out a set of cubes to each group (or get them to make up the cubes from the templates).

2. Ask the students to work out the missing number, by recording their observations and explaining how they got their answers.

3. At the end of the activity ask different groups to present their answers to the rest of the class.

4. Ask if other groups have got the same answer by a different route.

RS•C

5. Give the correct answers if no one has got them right.

6. This could then lead to a class discussion on the creativity involved in interpreting results.

7. When they have finished, get the students to make up their own cubes from the blank cube on the template and give them to other groups to solve. You could suggest that their cubes used letters as well as numbers.

Answers

There may be other valid patterns not included in the table.

Cube	Answer	Possible pattern
1	6	Consecutive numbers 1–6 Or opposite sides of the cubes add up to 7 Or numbers are arranged in the same pattern as found on a dice
2	8	Opposite sides of the cube add up to 10
3	36	The numbers on the cube are the squares of numbers 1–6

The cube activity

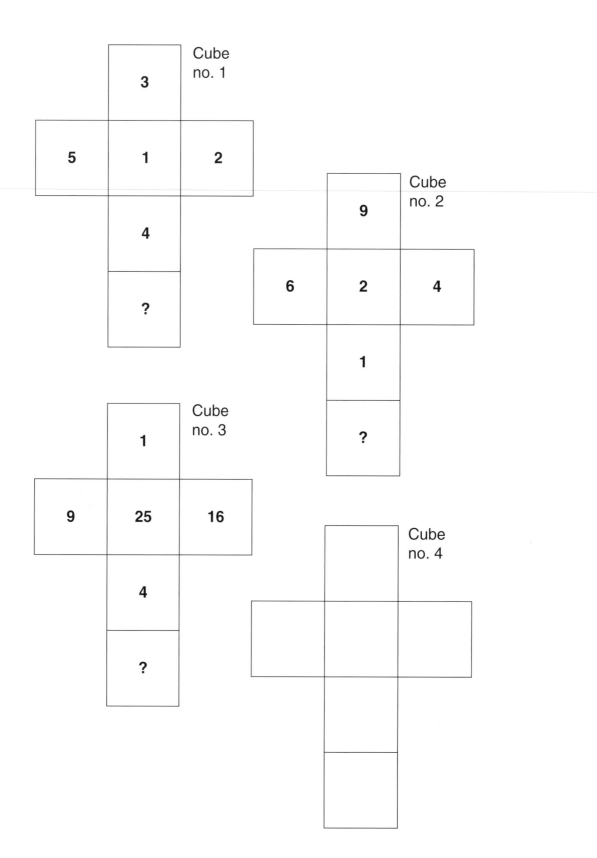

Cube no. 1

	3	
5	1	2
	4	
	?	

Cube no. 2

	9	
6	2	4
	1	
	?	

Cube no. 3

	1	
9	25	16
	4	
	?	

Cube no. 4

RS•C

Black box activity 3: A model tube

Teachers' notes

Objective
■ To understand the concept of modelling.

Outline
Black box investigations are a good way for students to learn about scientific theory and modelling because as the name suggests they are working in the dark. In real life scientists spend a lot of time working in the dark, trying to piece together data collected from various sources.

Teaching topics

This activity is appropriate for 11–14 year olds and could be used at any point in the scheme of work because no specific knowledge is required. However, it does lead nicely into thinking about a scientific model such as the particle theory of matter or the atom.

Resources

■ 1 cardboard tube about 30 cm long (*eg* from the centre of kitchen roll)

■ 1 plastic ring (optional, you can simply loop the lower rope over the upper rope)

■ Rubber bungs or tape (to seal the ends of the cardboard tube)

■ String.

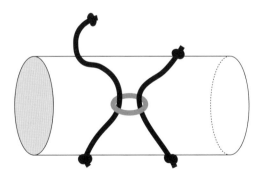

Figure 2 The inside of the model tube

For each group
■ Toilet roll tubes

■ String, tapes

■ Plastic rings optional.

RS•C

Timing

60–70 minutes

Possible lesson plan

1. Demonstrate the model tube to the class, pulling on different strings. Each time one string is pulled another will move. The pattern of movement will look quite complex. Tell the class that they are going to make a model of the tube, but they are not allowed to look inside.

2. Ask the class to observe carefully as the first string is pulled.

3. Ask the class to 'hypothesise' as to what is happening inside the tube.

4. Each member of the class should now draw a rough model of the inside of the tube.

5. Carry out a second test, by pulling a different string.

6. Ask the class to make another hypothesis.

7. According to the new evidence and hypothesis, the rough diagram should now be modified.

8. This process can be repeated until you think that the students have a good idea of what is happening in the inside of the tube.

9. Ask the class to make their models.

10. Gather the class together to compare the student's models with the original one. The discussion that follows will depend upon the quality of the models. You should aim to talk about the different models that are offered, about difficulties encountered, maybe frustrations experienced from not being allowed to look inside and stress that a model can never be the real thing.

11. The process of making a model should then be linked to science. You may wish to formally go through the process they have just carried out (see flow chart **Making a model or theory** in the introductory notes). Find out if certain students were influenced by what other people in the room were doing. Was there collaboration? Did some people give up on their own ideas and team up with others? Point out that real scientists often start out with their own ideas, but often the real progress is made when they exchange ideas and theories with other scientists. Often scientists team up and work together.

RS•C

Measurement, accuracy and precision

Teachers' notes

Objectives

■ Understand that data obtained during experiments are subject to uncertainty.

■ Understand that the level of accuracy is linked to the context.

■ Planning experiments and investigations.

■ Making accurate observations.

■ Evaluating data, considering anomalous results.

Outline

The teaching material is divided into three sections, all of which focus on an aspect of accuracy and precision. Each activity stands alone and is independent of the other two.

■ Measuring uncertainties and reporting reliable results

■ Choosing and using equipment

■ Does being accurate really matter?

Teaching topics

The activities can be used at any point in a course to teach investigative skills and are suited to students in the 11–16 age range. The activities can be adapted easily to allow access to students of different ages and of different abilities. In the Measuring uncertainties and reporting reliable results section **The weighing experiment (method 2)** could be modified and used as a post-16 key skills exercise. Although the activities are quite general in nature, they can be used to teach specific skills. Alternatively, they could be used prior to carrying out investigations that require weighing out of materials, or the measurement of volume and temperature.

For example **Choosing and using equipment** and **Does being accurate really matter?** could be taught prior to or after carrying out experiments or investigations into:

■ The neutralisation reaction between acid and alkali

■ Rates of reaction

■ Electrolysis

■ Methods of separation.

Experimental details can be found in *Classic Chemistry Experiments*[1] Numbers 48; 29, 64, 65; 81, 82; 1, 4, 71 and 100 respectively.

RS•C

Section 1: Measuring uncertainties and reporting reliable results

Background information
Students (including post-16 students) are often confused about the meanings and difference between some of the vocabulary in regular use *eg*

- Accuracy and precision

- Repeatability and reproducibility

- Systematic error and true value

- Error and mistake

- Best fit line and anomalous points.

Definitions
Accurate – the result is close to a reference value or the average of the data is close to a reference value.

Precise – the data points are close together (but there can be a random error).

Repeatability – when the experiment is repeated by the same person, using the same equipment and the results are close together.

Reproducibility – when the experiment is carried out by different persons, using different equipment and the results are close together.

True value – a perfect value of the quantity, *eg* mass, volume, temperature. This is an ideal and can never be known exactly.

Reference value – A value taken to be very close to the true value and usually accepted as a point of reference, *eg* a 'standard weight' has been measured on a balance that has little or no error and so the 'measured weight' is very close to the true value and accepted.

Errors are not the same as mistakes *eg* not reading a scale correctly.

Systematic error – there is some problem with the apparatus, because the results are precise (close together), but not accurate.

Instrumental errors – *ie* quantifying the precision of measurements. For example a 2 decimal place balance is precise to ±0.005 g.

Percentage errors – using a 2 decimal place balance, the errors when weighing 0.1 g and 0.01 g are identical yet the overall percentage errors are ±10% and 100% respectively (including the precision of the zero readings).

Overall percentage error – this arises when several measurements are used to achieve an overall result. It is approximately equal to the sum of the percentage errors in the individual readings although there are more complex treatments.
– In a simple weighing the overall percentage error is based on two readings, a zero or tare and the mass of the sample.
– In a titration the overall percentage instrumental error might be the sum of the percentage errors from the weighing, the volumetric flask, the pipette, the burette and the standard solution.

Reliability – this is assessed through comparison of an individual result with a reference or class mean.

Assessing the reliability of an individual result allows a judgement to be made regarding the level of mistakes. The overall percentage error based on instrumental readings is unavoidable whereas mistakes (human errors) are avoidable through repeating and reproducing the results.

Anomalous point – a data point that does not fit the pattern of the graph.

Trueness

Trueness is defined as: The difference between the observed mean value and the reference value.

A true value is never achievable because there is always some random variation and it is recommended that the indefinite article is always used *ie* 'a' and not 'the'.

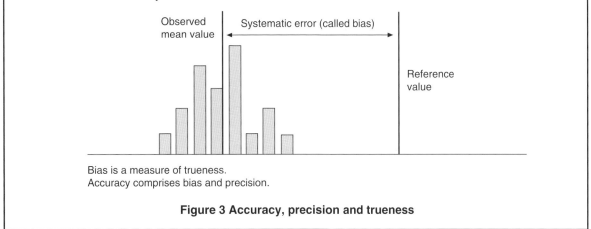

Bias is a measure of trueness.
Accuracy comprises bias and precision.

Figure 3 Accuracy, precision and trueness

Teaching tips

Introducing the vocabulary to the class
Professional judgement should be used here to decide how many terms it is appropriate to introduce to each class. The list of definitions given above is meant to be a comprehensive list to aid teaching at all levels.

There are several ways of introducing the terms to the class:

- List some of the words on the board and ask the students to write down the meanings. Discuss their answers.

- Use everyday ideas to introduce the terms and promote discussion *eg*
  – Accuracy and precision are required to succeed at darts and archery.
  – A cookery book must contain recipes that are repeatable and reproducible, otherwise no one would want to buy it.

- Using the student worksheet **Bulls eye to win**. This could also be used as a homework exercise.

- Analysis and interpretation of experimental data. The class could be presented with a set of results if there is not enough time to carry out the weighing experiment first.

RS•C

The weighing experiment (method 1)

- Set up the balances at different places in the room.

- Divide the class into groups.

- Present the class with some common identical objects – *eg* Mars® Bars.

- Each group weighs several Mars® Bars, using one weighing device.

- They then record the results and the weighing device used.

- They work out the average mass of a Mars® Bar.

- Add the result to the class table or graph (on the board or OHP).

- Plot the class results (from the table) individually.

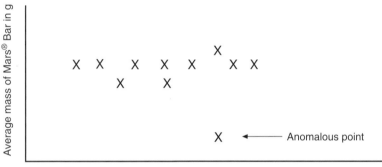

Figure 4 Sketch graph showing simple results

Interpreting the results

The results show a scatter close to a true value of the object (see Figure 4). You need to draw in the line of best fit and highlight any anomalous results. This approach is built on the assumption that all the Mars® Bars are the same mass, and that balances are the only variable. Any difference in the mass of the actual Mars® Bars will be minimised, because each group has used the average mass of several Mars® Bars in their result.

The weighing experiment (method 2)

As homework, ask the class to weigh the same object, such as a Mars® Bar, three times on their kitchen scales.

- Record the mass of the object each time they weigh it.

- Work out the average mass.

- Bring the object into school.

- Reweigh the object using a balance at school.

- Add the result to the class table or graph (on the board or OHP).

- Plot the class results (from the table) individually.

Interpreting the results

The results should show a scatter close to the declared value of the mass of the object. Manufacturers are given a tolerance on their products and therefore there is no true value or reference value. (Refer to notes in background information.) There will be a scatter of results, because in practice not all Mars® Bars will weigh exactly the same. You need to draw in the declared value (on the packet), allowed variability and highlight any anomalous results as shown in Figure 5.

RS•C

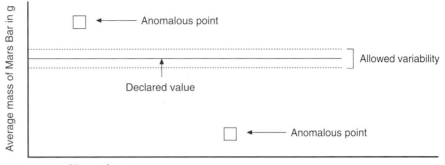

Figure 5 Sketch graph showing results, allowing for the variability in the mass of the Mars® Bar

The graph should be discussed either in groups or as a whole class. This could then be followed up by giving the students the worksheet **Interpreting the weighing experiment** either as a class exercise or a homework exercise.

Calibrating equipment

This should take the form of a teacher demonstration and discussion, and could be used in the final stage of the lesson. Some teachers may want their students to carry this out for themselves because it really depends on the quantity and quality of your balances. If available you could use a range of balances that read to a different number of decimal places. You may wish to demonstrate this to small groups at a time.

The class weighing exercise above should have already highlighted the fact that not all balances give an accurate reading, even if the readings are precise. It should be pointed out that from time to time, all balances need calibrating or resetting to make sure that they do give an accurate answer. An analytical chemist, carrying out measurements on a microscale (eg weighing to 0.001 mg) may have to calibrate the balance each time they use it. Even airflow can upset very sensitive balances as can the temperature of the object being weighed. The level of accuracy required is usually dictated by the application.

Resources

Classwork
- Mars® Bars or another product to weigh

- Kitchen scales (different types if available)

- Top pan balances of differing sensitivities

- Graph paper or graph plotting software or OHP.

- Student worksheets
  - Bulls eye to win
  - Interpreting the weighing experiment (method 1)
  - Interpreting the weighing experiment (method 2)

Demonstration
- Object of known mass such as a 10 g or 100 g weight

- 3 balances (if possible of different sensitivities). The least sensitive balance should not be calibrated correctly, as this will be done during the lesson.

RS•C

Practical tips

Class experiment

You may wish to add in an anomalous result, which could be used to discuss systematic errors later in the lesson. This could also save students from embarrassment, if there was only one other anomalous result. But in practice kitchen scales often provide the anomalous results. This could be checked before the lesson.

Demonstration

1. Using an object of known mass such as a 10 g or 100 g weight, place it in different positions on the most sensitive balance you have and record the results on the board. Dependent on the balance the reading will vary. How close is the reading to a 'true' value?

    Potential problem – make sure that you know a true value of your object at the start of the demonstration. Do not assume a 10 g weight has a mass of 10 g without checking it. It could be 10.1 g or 9.9 g.

2. Place the weight on a second balance (of different sensitivity). Do you get the same result?

3. Place the weight on a third balance and reweigh it. This time make sure that the balance is not zeroed correctly. If the balance has a setting-up levelling bubble, it should be off centre. The reading should be quite a bit out.

4. Show the group how to centre the bubble and carefully adjust the balance.

5. Re-weigh the object. It should now indicate the correct mass.

6. Stress the importance of correct setting up of equipment and of calibrating equipment, especially if accurate readings are required.

7. This leads naturally into a discussion about situations when accuracy is important and when it is not important.

8. The student worksheet **Does being accurate really matter?** could provide extension work or a follow up piece of homework.

Timing

2 hours (divided between classwork and homework)

Adapting resources

The student sheet **Bulls eye to win** could be adapted to meet the needs of the less able by turning question 3 into a cut and stick exercise and omitting questions 4 and 5.

Opportunities for other key skills

■ Application of number

Background information

Industrial trading standards

Industry must comply to the Weights and Measures Act of 1985 and the Weights and Measures (Packaged Goods) Regulations 1986.

RS•C

RS•C

In general terms the Act states: Goods which are sold in packages by weight or measure can be packed either to minimum quantity or to average quantity.[2]

For the minimum quantity each pack must contain at least the quantity marked on the pack (the nominal quantity). If equipment is used to make up the packs then the equipment must be tested and approved for trade use. The equipment does not have to be used, but if the quantity is estimated incorrectly, then the industry will have no defence.

For average quantity there are certain rules which must be followed called the Packers' Rules which are regularly monitored.

Packers' Rules

- The average content of the group must on average be at least the nominal quantity.

- No more than 2.5% (1 in 40) of the group may be non-standard ie (the nominal quantity) – (tolerable negative error).

- No package in the group may be inadequate ie (the nominal quantity) – 2 (tolerable negative error).

The tolerable negative error (TNE) is dependent on the nominal quantity.

Nominal quantity g or cm^3	Tolerable negative error
5–50	9% of nominal quantity
50–100	4.5 g or cm^3
100–200	4.5% of nominal quantity
200–300	9 g or cm^3
300–500	3.5% of nominal quantity
500–1000	15 g or cm^3
1000–10000	1.5% of nominal quantity
10000–150000	150 g or cm^3
Above 15000	1% of nominal quantity

Table 1 Legal industrial tolerance levels

The mass of the Mars® Bar

Teachers need to be aware of the common use of the word 'weigh' to determine mass. The average mass of a batch of products should be no less than the declared mass of 65 g.

The mass stated on the pack does not include the mass of the packaging.

Millions of Mars® Bars are produced daily. The few Mars® Bars weighed in class may not be representative of the batch.

Confectionery items weighing less than 50 g are not legally required to show a mass on the wrapper and this exemption applies to both standard and promotion packs.

RS•C

Smarties and Milky Bar Buttons

These are packed by mass rather than by the number of sweets. The manufacturing process is a fast and highly automated one and it is impossible to pack exactly the same quantity into each pack. As with any average there will be some packs with above average mass and some below. All packs will, however, be above an agreed minimum level.

Sources of information

http://www.tradingstandards.gov.uk/ (accessed June 2001)

Answers

Bulls eye to win

1. Jamal

2. The bulls eye

3. a) David, b) Helen, c) Jamal, d) Marie

4. David

5. Try an aim a bit lower and further over towards the right.

The weighing experiment (method 1)

1. Read off the value of the best fit line.

2. Use the value on the packet.

3 & 4 Accept sensible answers.

5. Mass of object (from graph) +/– difference (calculated in 4).

6. Accuracy of scales, precision of experiment, age of the product. Remember the manufacturer is allowed some variability.

7. See graph.

8. Dependent on results.

9. The balance has an error or the average was worked out incorrectly, or the individual results were not precise.

10. Accuracy is important; the customer should be getting the amount of product they are paying for.

The weighing experiment (method 2)

Questions 1–7 as method 1 above

8. To obtain the raw data of the anomalous result, go to the graph and find out whose measurement is incorrect. Go back to their exercise book and check the original readings. Write these readings up on the board. This should show if:
   (a) the average has been worked out incorrectly
   (b) if the results are not very precise
   (c) if there is a systematic error in the scales.

9. Dependent on results.

10&11 Refer to the raw data of the anomalous result.

12. Dependent on results.

RS•C

13. See reasons given under 8.

14. Make sure that you know how to work out averages, each time the object is weighed make sure that the balance reads zero and that nothing has been spilt on it *etc*.

Bulls eye to win

Jamal, David, Marie and Helen spent the afternoon playing darts. In the last round they set the target as the bulls eye. Each person was allowed seven throws. The results of their game are shown below.

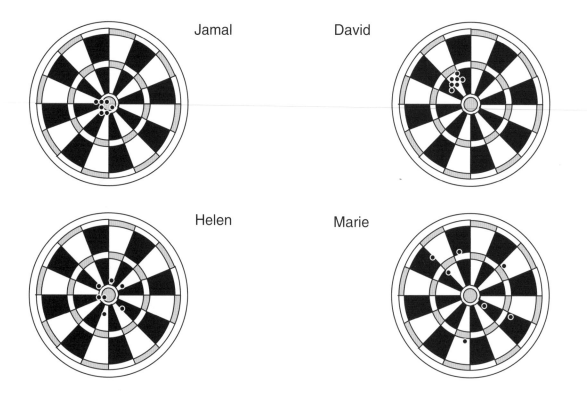

Jamal David

Helen Marie

Questions

1. Who won the game of darts?

2. What did the players choose to have as their reference value?

3. Whose game would you describe as:
   (a) Precise but inaccurate
   (b) Imprecise but accurate
   (c) Precise and accurate
   (d) Imprecise and inaccurate?

4. Who do you think needs to improve their game to avoid experiencing the same systematic error next time they play?

5. What advice would you give that person?

Interpreting the weighing experiment (method 1)

1. From the graph produced what is the mass of the Mars® Bar?

2. What is the expected value of the mass of the Mars® Bar? (Look at the wrapper.)

3. Is there a difference between the measured value and the expected value?

4. What is the size of the difference?

5. How would you write down the mass to include this error?

6. How can you explain the difference in results?

7. Are there any anomalous results?

8. Are the anomalous results showing a systematic or random error?

9. Suggest a reason to explain the anomalous result.

10. Discuss the importance of accuracy in the manufacture of Mars® Bars.

Interpreting the weighing experiment (method 2)

1. From the graph what is the mass of the object?

2. What is the expected value of the mass of the object? (Look at the wrapper.)

3. Is there a difference between the measured value and the expected value?

4. What is the size of the difference?

5. How would you write down the mass to include this error?

6. How can you explain the difference in results?

7. Are there any anomalous results?

8. If the answer is yes to question 7, check and record the raw data of the anomalous result.

9. Are the anomalous results showing a systematic or random error?

10. If the precision is poor, is it an example of poor repeatability or poor reproducibility?

11. If the results are biased (systematic error) what could this cause?

12. For the anomalous result, is there a difference between the home result and the school result? If so, what?

13. Suggest a reason to explain the anomalous result.

14. Suggest how this problem could be avoided in the future.

RS•C

RS•C

Section 2: Choosing and using the right equipment

Teachers' notes

Many students have difficulty choosing the best apparatus to carry out the experiments and investigations. The worksheet **Choosing and using equipment** is designed to help in this.

Teaching tips

For lower ability students, supply each group with the actual equipment. The students should then be able to experience the different scales. For example, they could fill each measuring cylinder or beaker with 30 cm^3 of water, and use their observations to help decide upon the correct piece of equipment.

For more able students this activity could be introduced in class and then carried out as a homework exercise.

Resources

■ Student worksheet
– Choosing and using equipment

Timing

60 minutes or one lesson

Answers

1. At this stage the student is not expected to give the size of the beaker, measuring cylinder or thermometer, however they should include them in a list or a labelled diagram showing a heatproof mat, Bunsen burner, tripod and gauze.

2. C = 50 cm^3, there is no 30 mark on A, B is too small, the divisions on the scale on D go up in 10s whereas they go up in 5s on C. So more accurate to use C.

3. C. Water is usually heated in beakers. B and D are not beakers. More water will evaporate away if A is used as the water has a much larger surface area.

4. A or E. B and D do not go up to 100 °C, and C will not be so accurate as the same size thermometer goes up to 200 °C. E may or may not be the most accurate/easy to use.

5. Group 1 = 90.5 °C, group 2 = 98.6 °C, group 3 = 99.5 °C, group 4 = 101.2 °C

6. Group 3

7. Group 3

8. Groups 1 or 3

9. The result is much lower than expected.

10. Any 2 of:
    (a) James could have used the wrong sample of liquid.
    (b) His technique may not be very good. He may be taking the thermometer out of the water to read the scale.
    (c) The thermometer could have a systematic error.

11. The appropriate pair of answers from:
    (a) Repeat using the correct water supply.
    (b) Repeat, making sure that the thermometer is not taken out of the water.
    (c) Repeat, using a different thermometer, (or borrow results from group 3).

Choosing and using equipment

As part of a class investigation, James has been asked to heat 30 cm^3 of water and measure the temperature at which it boils. From his knowledge of science, he knows that water boils at 100 °C at a pressure of 1 atmosphere.

When James opened the cupboard to get out the equipment, he was very surprised to see so many different sized beakers, conical flasks, measuring cylinders and even thermometers. 'What should I use to get the most accurate result?' thought James.

After choosing the equipment and taking some water from the container labelled 'distilled water', he carried out the experiment and noted down the result. However, he was not happy with the first result, so he repeated the experiment. He recorded his results in a table.

Your job is to help James choose the correct equipment.

1. List all the pieces of equipment James will need to use to boil 30 cm^3 of water.

2. Which of the following containers should he use to measure out the water? Give a reason to support your answer.

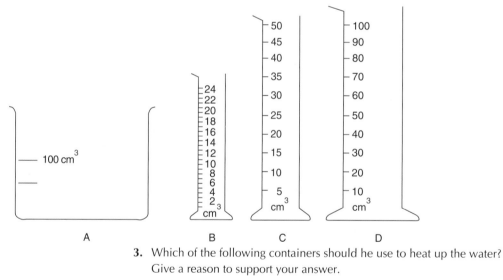

3. Which of the following containers should he use to heat up the water? Give a reason to support your answer.

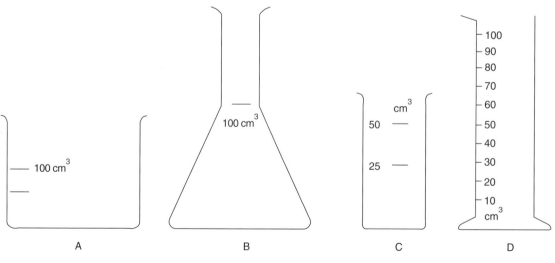

4. Which of the following thermometers should he use to measure the temperature of the water?
Give a reason to support your answer.

A	B	C	D	E
°C	°C	°C	°C	
110	60	200	70	
100	55	180	60	
90	50	160	50	
80	45	140	40	
70	40	120	30	
60	35	100	20	23.1 °C
50	30	80	10	
40	25	60	0	
30	20	40	-10	Temperature range
20	15	20	-20	-50 °C to 500 °C

The table shown below shows the results of the groups in James' class.

| Test | Boiling temperature / °C | | | |
	Group 1 & James	Group 2	Group 3	Group 4
1	90.0	95.0	99.0	102.5
2	91.0	100.0	100.0	100.0
3	90.5	101.0	99.5	101
Average				

5. Work out the average temperature for each set of results.

6. Which group has measured an average boiling temperature closest to the expected value, if the pressure is 1 atmosphere?

7. Which group has the most accurate result?

8. Which group has the most precise readings?

9. Why do you think James is not happy with his group's result?

10. Suggest two reasons why James' results are different to the rest of the class.

11. Suggest how James could test out two reasons.

RS•C

Section 3: Does being accurate really matter?

Background information

This activity is intended to help pupils appreciate that different situations in science and everyday life require different degrees of accuracy and therefore different instruments to carry out the measurements. For example, if someone were intending to break the world record in the 5000 m run, they would need a stopwatch reading to 0.01 s during training, but if they just wanted to keep fit a normal wristwatch would do.

During accurate work, there is a need for instrument calibration. Precise results do not necessarily mean accurate results, if there is a systematic error. In some situations a systematic error will be negligible with no overall real effect, but in others it could be the difference between life and death if for example, the pharmacist's balance has a 15% error. It could become very expensive if the balance used to weigh gold ingots was incorrect while, if quality control in Mars® bar manufacture had a systematic error we might be lucky and get a bigger Mars® bar.

It should also be stressed that a systematic error is not important if the weighing is done by difference.

Teaching tips

Introduce the lesson using everyday examples of when measurements need to be accurate. You could go through some of the examples on the worksheet to get the class started. This lesson is suited to group work or individual work. Some of the answers to question 1 are totally dependent on context.

Resources

■ Student worksheet
– Does being accurate really matter?

Timing

60–70 minutes

Adapting resources

This worksheet can be adapted to meet the needs of the less able or younger student by:

■ Reducing the number of statements on the sheet.

■ Enlarging the statements in question 1, photocopying them on to card and then cutting out the statements so that the students can physically sort them into two groups.

■ Replacing the written questions with pictures and then proceeding as above.

■ Using a cut and stick approach to question 1 to minimise the writing.

■ Reduce the difficulty of question 2, by only asking for one variable to be found.

■ Removing questions 3 and 4 which are more demanding.

This worksheet can be adapted to meet the needs of the more able or older student by:

■ Asking the students to add in some more situations.

RS•C

- Add the following to question 4. Suggest the degree of precision the instrument should have.

- Giving the student some numerical examples to work with *eg*
  - If a premature baby weighing 2 kg loses 0.5 kg; what is the overall percentage reduction in body mass?
  - If an adult man weighing 80 kg loses 0.5 kg; what is the overall percentage reduction in body mass?
  - Using your answers explain to what degree of accuracy you would measure the mass of the premature baby and the adult man.

Opportunities for other key skills

- Application of number

- Problem solving

- Working together.

Answers

1. Note that two of the measurements may be put into both categories.

Accurate measurements	Rough estimations
Timing 100 m breast stroke race.	Making a drink of squash.
Weighing a premature baby – a mass loss of 500 g could be vital. **(C)**	Weighing an adult man. A measurement to the nearest kg will do.
Running in an international 800 m race.	Going for a 3 km jog each morning – it doesn't matter how far you go or how long you take as it is only for fitness.
Training for a 800 m race – practice of the real thing.	Training for a 800 m race – different distances may be run to gain general fitness etc.
Building an Olympic swimming pool – if the length is slightly wrong (even by 1 cm) then new Olympic or World records cannot be set. **(C)**	Measuring out sand, cement and gravel for concrete – this is done by parts, where a part could be a bucket full or a shovel.
Weighing out aspirin to make it into 500 mg tablets – a small difference in weight could have a dramatic effect on the patient. The tolerance level is +/– 15%. **(A)**	Measuring the air temperature.
Analysis of water – the volume of water must be known accurately so that the concentration of the pollutants can be measured. **(B)**	
Weighing an adult man if he is a boxer or jockey.	
In an orienteering competition, the direction is measured using a compass.	In an orienteering competition, the distance is often estimated by pacing.
The body temperature of a child with a fever. **(B)**	

RS•C

2.

	Variable 1	Variable 2
100 m breast stroke	Timing (**C**)	100 m length (**C**)
Running a 800 m race	Timing (**C**)	800 m distance (**C**)
Training for a 800 m race	Timing (**A**)	Distance (**A**)
Orienteering competition	Direction (**A**)	Distance (**A**)

All the other situations only have one variable that needs measuring.

3. See letters in bold, in the above tables.

4. Any acceptable instruments. In this answer it is important for the degree of precision to be given with the measuring instrument.

Does being accurate really matter?

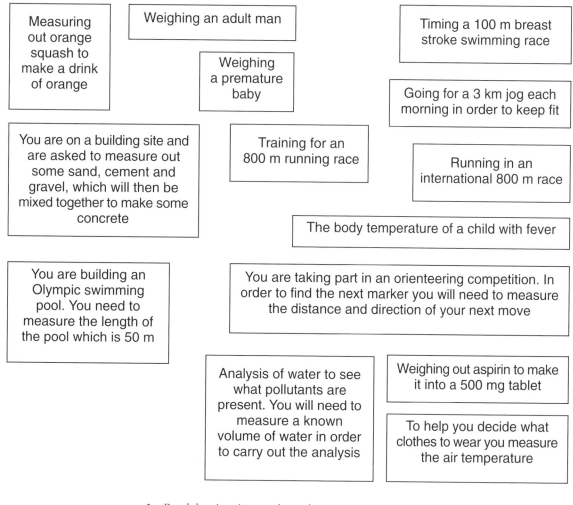

Measuring out orange squash to make a drink of orange

Weighing an adult man

Weighing a premature baby

Timing a 100 m breast stroke swimming race

Going for a 3 km jog each morning in order to keep fit

You are on a building site and are asked to measure out some sand, cement and gravel, which will then be mixed together to make some concrete

Training for an 800 m running race

Running in an international 800 m race

The body temperature of a child with fever

You are building an Olympic swimming pool. You need to measure the length of the pool which is 50 m

You are taking part in an orienteering competition. In order to find the next marker you will need to measure the distance and direction of your next move

Analysis of water to see what pollutants are present. You will need to measure a known volume of water in order to carry out the analysis

Weighing out aspirin to make it into a 500 mg tablet

To help you decide what clothes to wear you measure the air temperature

1. Read the situations and sort them into two groups. The first group requires accurate measurements. In the second group a rougher estimation will be good enough. Present your answer in a table.

2. For each of the situations that require accurate measurements, decide if there is more than one variable that needs to be accurately measured. Record your results in a table. The first one has been done for you.

	Variable 1	Variable 2
100 m breast stroke	Timing	100 m length

3. For each of the situations that require accurate measurements decide how accurate each reading must be? Use the code A – accurate to 1 unit, B – accurate to 1/10 of a unit and C – accurate to 1/100 of a unit and add these letters to your table.

4. Name the type of instrument that you would use to make each measurement and give its precision.

RS•C

Scurvy –
The mystery disease

Teachers' notes

Objectives
■ To understand the scientific method and how it has developed over the years.

■ To learn about how scientists worked in the past.

■ To understand that vitamins are essential for a healthy body.

Outline
This activity only requires a brief introduction and then the class can work thought it independently. It should be emphasised that early scientific theory depended on careful observation fitting in with the common theories. No experiments were carried out. Modern scientific thinking requires theories to be checked out by carefully designed experiments. In this piece of work it is possible to see how scientific method developed over 400 years.

Teaching topics

This activity is suitable for 14–16 year olds. The students will need the following pre-knowledge:

■ Oranges and lemons are acidic

■ Acids react with alkali to make neutral substances

■ The meaning of a control experiment.

It could be included in a unit on acids and alkali or as an introduction to investigating which foods contain vitamin C.

Background information

Science, technology and medicine
Throughout history, the study of medicine has been considered to be very important. 5000 years ago many of the medicines were based on common sense ideas. However, people also used charms and spells to keep away evil spirits that were thought to cause disease and suffering. Slowly over time a more scientific approach was adopted. This is shown in Figure 6.

RS•C

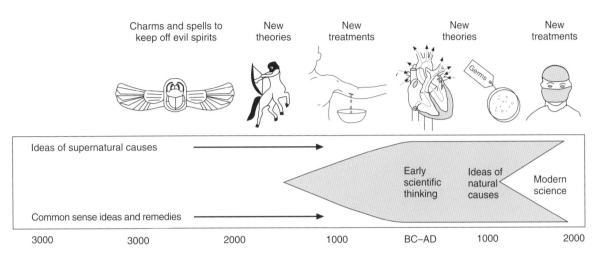

Figure 6 The development of scientific method
(Adapted and reproduced with permission from J. Scott, C. Culpin,
Medicine through time, Collins Educational, 1996.)

In the 16th century, 'modern science' began. Since then scientists have used it to build up, for example, a detailed knowledge of the workings of the body. They carefully test and check each part of the knowledge so that they can apply the knowledge to problem solving. It is really important for doctors to understand scientific methods so that research can continue.

The diagrams in Figure 6 explain the main difference between early scientific thinking and modern scientific thinking. The change from one to another was quite slow. Some parts of scientific method had been used for a long time, whereas others were not fully understood until the 19th century.

Suggested question

What were the main differences between the two sorts of thinking?

RS•C

Figure 7 Scientific thinking
(Adapted and reproduced with permission from J. Scott, C. Culpin, *Medicine through time*, Collins Educational, 1996.)

Teaching tips

Approach 1

The students work through the student worksheets, answering the questions.

Approach 2

The students carry out the James Lind role-play. This is an opportunity to include some aspects of the teaching of citizenship. The role-play can be treated as a stand-alone lesson without carrying out the other worksheets.

The James Lind role-play

This should be done before the class have seen the results of James Lind's experiment, otherwise it might influence what the sailors say and the decisions that are made.

Working in groups, the students are to play the role of James Lind and the sailors who are suffering from scurvy. James Lind should have an assistant or two to help him make the difficult decisions.

- Give a 5 minute introduction to the lesson explaining who James Lind was, what the scurvy problem was and how for about 200 years it had been known that eating fresh oranges, lemons and some other foods cured scurvy.

- Divide the class into groups of between 6 and 8.

- Sub-divide the group into sailors and James Lind + assistants.

- Give out the task sheets.

- Allow between 10 and 15 minutes for the sub-groups to read and discuss their task and start to think about possible strategies.

- Give the groups 15 minutes to carry out their role-play.

- One person from each group should report back to the rest of the class about how they decided which sailors would take part in the investigation and who got which treatment. You may find that some groups did not come to a conclusion.

- During the class discussion you need to get the groups to say how they tackled the problem, (eg looking for sailors with similar build etc. to make the test as fair as possible, consideration of the family background of individual sailors).

- Interviewing James Lind and the sailors is a useful technique for promoting discussion. For example, James Lind could be asked the following questions:
  – How did you decide which sailors to treat?
  – How did you decide which treatments to give to which sailors when you knew that some would work and others may not?
  – Were you worried about the effect some of the treatments could have on the sailors?

  The sailors could be asked the following questions:
  – How did you feel when you first heard about the experiment?
  – What do you think of James Lind?
  – How do you react to seeing other people getting better when you are not?

- The role-play should be followed up by revealing the results of Lind's experiment. This could be done either at the end of the lesson or during the next lesson.

- The class will now be in the position to answer the questions on student worksheet **Scurvy – the mystery disease**. This could prove to be a useful homework exercise.

RS•C

Resources

■ Student worksheets
  – Scurvy – the mystery disease
  – Task sheet for the sailors suffering from scurvy
  – Task sheet for James Lind and his assistants.

Timing

Approach 1 A 60–70 minute lesson and homework.

Approach 2 A 60–70 minute lesson for the role-play, with a second lesson needed to complete the other sections.

Answers

1500s

1. Explanation 1 fits the theory 'bad air carries disease'. After 12 weeks they had got to a place with bad air. Explanation 2 fits the four humours medical theory.

2. Conditions at sea would have been cold, the sailors became gloomy, depressed and melancholy. It was suggested that they should get more exercise and so the hornpipe dancing was introduced.

3. The four humours theory did not encourage tests and experiments. Medicines were given on a trial and error basis.

4. Acidic medicines such as orange and lemon juice cured scurvy.

5. See if any non-acidic fruits or herbs cured scurvy.

1600s

1. By carrying out careful measurements Sanctorius discovered that people lose and gain weight all the time, regardless of the balance of the four humours.

   Sanctorius knew that he caught the skin disease from a glove and not because the balance of the four humours was wrong.

1700s

1. Lind thought scurvy was an alkaline disease.

2. Yes, group 4 were not given an acid but seawater.

3. Yes, they all had the same menu. Only the treatment was varied, but he did run out of oranges after a week.

4. He needed more evidence to draw firm conclusions. There were only two people in each group and not all acids cured scurvy.

5. Have larger groups, run the experiment for a longer period of time.

6. Lind based his theory on the work of Sanctorius and the bad air disease.

7. Yes

8. Try out the experiment in warm, dry conditions.

9. The conditions were not the same, maybe the oranges and lemons were not very fresh.

RS•C

1900s

1. (a) Holst predicted that feeding guinea pigs on a diet of polished rice would give them beri-beri.

    (b) Holst observed the symptoms of scurvy rather than beri-beri.

    (c) Holst thought that giving the guinea pigs orange and lemon juice would cure the scurvy.

2. Guinea pigs are mammals and more like humans than birds.

3. Firstly they gave the guinea pigs scurvy by feeding them polished rice and then they fed them with one type of food such as cabbage or lemon juice to see if the scurvy was cured.

4. A deficiency disease is when you become ill because something is missing form your diet.

5. I would accept Holst's theory because all the experiments show that when certain foods are left out of the diet, the guinea pigs got scurvy and when they were replaced the guinea pigs got better. Boiled milk was the reason why the babies of the rich were getting scurvy in the 1800s. Holst's theory did not give any evidence that suggested a scurvy bacteria.

6. Vitamin C contains carbon, hydrogen and oxygen atoms.

7. Citrus fruits, green vegetables and potatoes all contain vitamin C.

Scurvy –
the mystery disease

Understanding scurvy, a problem for over 400 years, saw many changes in scientific thinking and method.

The story began in the 1500s

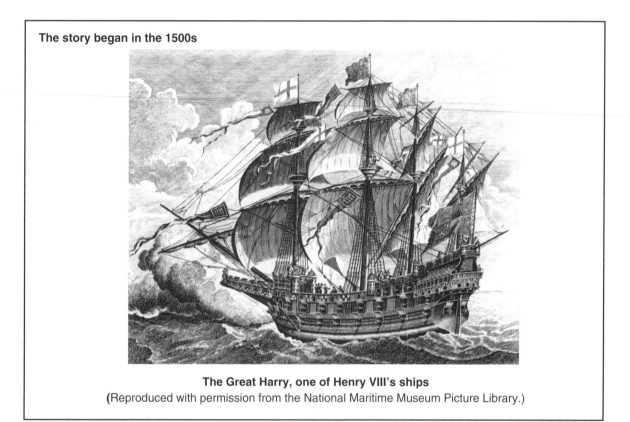

The Great Harry, one of Henry VIII's ships
(Reproduced with permission from the National Maritime Museum Picture Library.)

After 12 weeks at sea, many sailors got ill. Later the illness was called scurvy. Eating oranges, lemons and fresh food cured the sailors.

Common theories about disease

- Bad air carried disease

- You could get malaria by damp marshes

- You could get flu from being next to someone who had flu

- The four humours medical theory

- Healthy people had a good balance of the four humours.

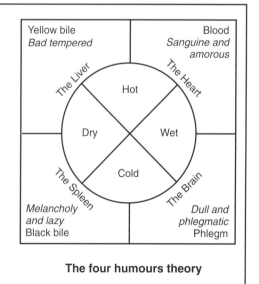

The four humours theory

Possible explanations for the sailor's illness

■ The air at that latitude was bad

■ Too much black bile

■ An alkaline disease.

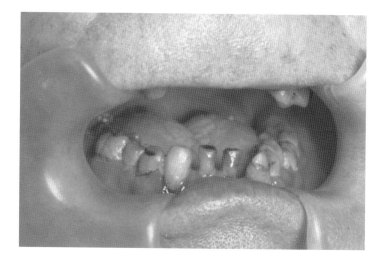

The effect of scurvy on the mouth
(Picture: Science Photo Library.)

Questions – 1500s

1. Did the explanations fit the common disease theories?

2. What evidence led doctors to believe there was too much bile?

3. Do you think there was a problem with the four humours theory?

4. What evidence led some doctors to believe scurvy was an alkaline disease?

5. How could they test out the alkaline theory?

The 1600s – a new scientific approach

■ Thermometers and balances had been invented and scientists were starting to carry out experiments by doing careful measurements.

■ After careful weighing experiments Sanctorious concluded that 'People lose weight by invisible perspiration through the pores on the skin'.

■ Sanctorious also caught a skin disease after picking up a glove from a lady who already had the skin disease.

Questions – 1600s

Give two reasons why you think that Sanctorious, along with some other scientists rejected the four humours theory.

1700s – Scotsman James Lind investigates Scurvy

While at sea, Lind gathered together 12 similar looking sailors, all suffering from scurvy. He then divided them into six pairs and gave them the following treatments.

Group No.	Treatment
1	2 pints of cider each day
2	A daily gargle with 25 drops of sulfuric acid in water
3	2 teaspoonfuls of vinegar three times a day
4	Half a pint of seawater a day
5	2 oranges and 1 lemon a day
6	A mixture of nutmeg, garlic, mustard, myrrh and radish root, plus barley water acidified with tamarinds.

Daily menu for a two week trial
Breakfast   Gruel (soaked bread) with sugar

Lunch   Mutton broth

Supper   Barley, raisins, rice and currants

Problem
After a week all the oranges had been eaten.

Results after 1 week
Group 5 were well.

Group 1 were getting better.

Groups 2,3,4,6 showed no improvement.

Results after 2 weeks
Group 1 were almost better.

Groups 2, 3, 4, 6 still showed no improvement.

Lind's interpretation of the results

Oranges and lemons contained a special substance. He called it 'Antiscorbutic'. He thought that the air in a cold wet climate might block up the important pores in the skin through which so much perspiration had to pass. Then the blocked perspiration went bad inside the body, causing scurvy. The oranges and lemons formed a kind of soap with the stale fat in the body which washed out the blocked pores and the scurvy was cured.

At home in Scotland, Lind tried to repeat his experiment, but did not get very good results.

Questions – 1700s

1. Which earlier theory do you think Lind based his experiments on? Explain your answer.

2. Do you think that Lind had a control experiment?

3. Did Lind carry out a fair test?

4. Do you think that he had enough evidence to draw firm conclusions?

5. How could he have got more evidence?

6. Which theories did Lind base his interpretation on?

7. Do you think Lind had a good imagination?

8. What experiment would you do to prove Lind's interpretation is correct?

9. Why do you think that Lind did not get very good results when he repeated the experiment at home?

1800s – Babies of the rich hit by disease

Symptoms
■ Sore bodies

■ Swollen and bleeding gums

■ Swollen legs

Observations
Age: 10–15 months old

Diet: Bread, butter, boiled milk.

1897 American doctors investigate and link this disease with scurvy. They conclude 'they are the same'. A diet of raw cows milk, orange juice and raw beef juice cures the babies.

Meanwhile... Chemists investigated acids, analysed orange, lemon and lime juices and found that they all contained 'citric acid'. Further experiments showed that citric acid was not the 'antiscorbutic' which cured scurvy; neither was boiled juice or concentrated juice effective.

1900s – The scurvy disease still not understood

A new model of disease is needed.

Norwegian scientist, Axel Holst finds the answer to scurvy while trying to solve the problem of beri-beri.

Holst's experiment
Holst knew from work carried out by the Dutch that chickens and humans that did not eat a certain substance that is contained in normal rice but not in cooked or polished rice, became ill with beri-beri. Beri-beri was not an infection, it was not a poison and it did not come from infected air.

At first Holst experimented on pigeons but then he changed to experiment on guinea pigs. He fed the guinea pigs on a diet of polished rice. The guinea pigs began to show more signs of scurvy than beri-beri. After 30 days of polished rice, the guinea pigs were fed lemon and orange juice. The guinea pigs got better.

Holst joins up with scientists studying scurvy
Experiments during the next few years showed that the following foodstuffs all stopped guinea pigs from getting scurvy:

■ fresh cabbage

■ lemon juice

■ apples

■ milk unless it was heated to 100 °C

■ sprouting grains and peas.

Conclusions about scurvy

- ■ It was not an infection
- ■ It was a deficiency disease
- ■ Later, the missing substance was called vitamin C.

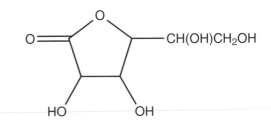

The vitamin C molecule

Response to 'The Holst Theory'

Some scientists were convinced and believed the 'deficiency' theory. Other scientists continued to look for 'scurvy bacteria'.

Questions

1. Which part of Holst's experiment would you describe as
   (a) prediction
   (b) observation
   (c) clear thinking?

2. Why do you think Holst decided to use guinea pigs instead of pigeons?

3. How do you think the scurvy scientists carried out their experiments with cabbage etc?

4. What is meant by a deficiency disease?

5. From the evidence presented, how would you respond to the 'Holst Theory'? Give reasons for your answer.

6. What elements are in the vitamin C molecule?

7. Which foods contain vitamin C?

Task sheet for the sailors suffering from scurvy

1. You have heard that James Lind is looking for a cure for scurvy. Apparently he wants some of you to take part in the experiment. What is your first reaction? Do you want to be involved?

2. The rumours say that some treatments are a lot nicer than others. Everyone knows that fresh oranges and lemons cure scurvy. If he uses oranges and lemons in his experiment then you will be cured. How can you make sure that you get the oranges?

3. How do you feel about using other treatments? No one knows if they will work.

4. James Lind has arranged a meeting for all the scurvy sufferers at 2 p.m. on the deck. What are you going to say to him, when he tells you about the investigation? Will you help or is the risk not worth it?

Task sheet for James Lind and his assistants

- You must tell the sailors that you are going to investigate scurvy. You think that scurvy might be an alkaline disease and so you intend to try different acidic treatments. It is well known that oranges and lemons cure scurvy.

- You must tell the sailors about how you intend to carry out the investigation. Details are given below for the two week trial. Each group consists of two sailors.

- You must decide which sailors are to take part in the experiment.

- You must decide which sailors are to be given each treatment.

Group No.	Treatment
1	2 pints of cider each day
2	A daily gargle with 25 drops of sulfuric acid in water
3	2 teaspoonfuls of vinegar three times a day
4	Half a pint of seawater a day
5	2 oranges and 1 lemon a day
6	A mixture of nutmeg, garlic, mustard, myrrh and radish root, plus barley water acidified with tamarinds.

Daily menu for a two week trial

Breakfast
Gruel (soaked bread) with sugar

Lunch
Mutton broth

Supper
Barley, raisins, rice and currants

RS•C

Concept cartoons

Teachers' notes

Objectives

- To promote class or group discussions about difficult scientific concepts.

- To introduce questions that students can investigate.

- To help understanding of experimental results and observations.

Outline

Two examples of concept cartoons have been included in this material

- Brewing up

- Flickering candles.

Teaching topics

This material is aimed at 11–14 year olds and could be included in the following topics

- Combustion

- Burning fuels.

Background information

Using concept cartoons to introduce investigations helps students to think through the scientific ideas associated with the question.

In each concept cartoon a scientific question is asked. The rest of the concept cartoon presents the student with alternative viewpoints or different 'theories' on scientific concepts that relate to the question that has been asked. The concept cartoons can be designed to have more than one correct idea. During the discussion all ideas presented on the concept cartoon are given equal weighting, thus promoting an ideal opportunity to discuss scientific concepts within a safe environment. Students should then feel more confident to put forward their own ideas.

Concept cartoons can be used to promote group discussion in other situations such as after demonstrating an experiment (eg Brewing up) or after the students have carried out their own experiment and are trying to make sense of their observations or data (eg Flickering candles).

Sources of information

There are further examples of concept cartoon in the following books;

B. Keogh and S. Naylor, *Starting Points for Science*, Cheshire; Millgate House Publishers, 1997.

S. Naylor and B. Keogh, *Concept Cartoons in Science Education*, Cheshire; Millgate House Publishers, 2000.

The CONCISE project provides training for teachers in the use of concept cartoons and details may be obtained from B. Keogh and S. Naylor, Institute of Education, Manchester Metropolitan University.

RS•C

Teaching tips

Using concept cartoons as investigation starters
- Present the concept cartoon to the class either by using photocopies or an OHP.

- Read out the question to the class.

- Go through all the possible answers given on the sheet.

- Ask if anyone has any other suggestions.

- Ask the class which they think is correct. If possible they should try and give a reason. At this stage accept all answers.

- Tell the class that they are now going to plan an investigation to see if their hypothesis is correct.

- Depending on the class and the amount of planning they have done before, structure the rest of the lesson as a normal planning lesson.

Using concept cartoons to promote group discussions
- Carry out a teacher demonstration or class experiment.

- Give each group a copy of the concept cartoon and ask them to discuss the ideas.

- They then write a conclusion to their experiment and where possible they must give reasons to support their answers.

Brewing up

The concept
When heating a container of water on a gas stove, students will initially observe water droplets on the cold side of the container. As the container heats up, the water droplets will disappear.

When hydrocarbons are burnt, water is one of the products. The water condenses out on the cold surface, but evaporates as the surface is heated. One of the common misconceptions held is that water vapour already present in the air condenses out.

The investigation
Students should carry out an investigation, making sure that the sides of the beaker are dry at the start to see if any water appears on the side of the beaker during the heating of water. They should plan to carry out a fair test, with the source of heat being the variable. You will need to have a range of energy sources available.

Students should only observe the water droplets when hydrocarbon based fuels are used, and not when electric heaters are used. This should allow them to work out the answer.

Further investigations could be carried out to verify that the liquid really was water, by testing it with anhydrous copper sulfate or cobalt chloride paper. The more alert student will observe that after a while the liquid disappears. An investigation could be carried out to determine the temperature at which this starts to occur.

As this in an open-ended investigation the following list is a guide for possible resources and not meant to be a definitive list.

RS•C

Resources (available for investigation)

- Bunsen burners

- Tripod

- Gauze

- Heat proof mat

- Electric hot plates

- Immersion heaters

- Camping gas burner

- Candles

- 250 cm^3 beaker

- 250 cm^3 metal or ceramic container (to heat the water in)

- 100 cm^3, 150 cm^3, 200 cm^3 measuring cylinders

- Thermometer

- Anhydrous copper sulfate

- Cobalt chloride paper

- Tongs

- Safety glasses

- Student worksheet
  – Brewing up

Flickering candles

The concept
Students are often puzzled and confused about what happens when a candle burns. It is quite common for the student to believe that only the wick burns and not the candle! The supporting evidence for this may come from observing liquid wax running down the side of the candle. This idea presents more room for confusion, as they try to work out how the wax became liquid. Was it by melting or dissolving? The other area of confusion lies in the distinction between evaporation and burning. Many students will fail to realise that, in the case of wax, the combustion products are gaseous.

The investigation
There is plenty of scope for investigating the combustion process. For example, students could carry out tests to distinguish between melting and dissolving or determine whether there was a mass change during burning. To distinguish between burning and evaporation students could collect the gases and test them by cooling then down to see if they condense out. If time permits, further tests for carbon dioxide and water should be carried out.

As this in an open-ended investigation the following list is a guide for possible resources and not meant to be a definitive list.

Resources (available for investigation)
- Candles

- Petri dishes

RS•C

- 250 cm^3 beakers or water containers
- 100 cm^3, 150 cm^3, 200 cm^3 measuring cylinders
- Balances
- Lime water
- Anhydrous copper sulfate
- Anhydrous cobalt chloride paper
- Droppers
- Tongs
- Safety glasses
- Student worksheet
  – Flickering candles

Timing

This really depends on how you want to use the rest of the lesson. 15 minutes should be allowed for the discussion.

A lesson and homework should be enough time to plan the investigation. Further lesson time would be needed to carry out the investigation and analyse the results.

If the concept cartoon is being used for an assessment at the end of a topic, then it should only take a few minutes for the students to work out the correct answer.

Adapting resources

To enable less able students to plan their own investigations, the following sheets could be produced:

- A list of experiments that the student could match up to the statements in the cartoon. They could then choose which experiment they were going to carry out.
- A list of prompt questions for the student to answer eg What equipment will I need? How can I make this a fair test?

Brewing up

Flickering candles

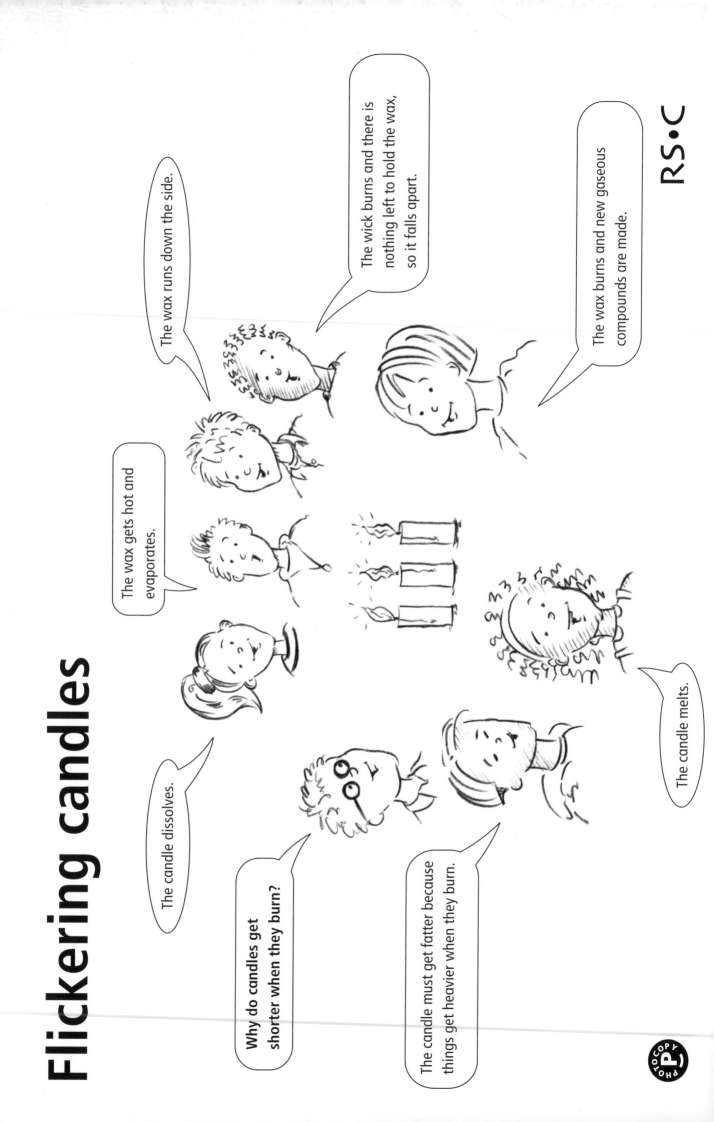

RS•C

References

1. K. Hutchings, *Classic Chemistry Experiments*, London: Royal Society of Chemistry, 2000.

2. Guidance Leaflet *Guidance Notes on average quantity*, Trading Standards, 1999.

3. W. F. McComas, *The Nature of Science in Science Education*, Dordrecht: Kluwer Academic Publishers, 1998.

4. J. Solomon, *Discovering the Cure for Scurvy*, Hatfield: Association of Science Education, 1998.

5. T. Lister, *Classic Chemistry Demonstrations*, London: Royal Society of Chemistry, 1995.